INTRODUCTION

DEDICATION

This book is dedicated to my family and friends, who, although extremely proud of me, are probably too squeamish to be able to open up this book, which is loaded with pictures of lizards and insects (and lizards eating insects). I presumably will have to convey this thought verbally to them, as they will never read it.

FOREWORD

The subject of feeding animals as large as the group known as reptiles could make for an expansive volume of literature. However, with limited space we will restrict our topic to insectivorous (insect-eating) lizard species only. These are generally small to medium-sized lizards that inhabit a multitude of habitats, from swamp to desert. The majority of lizard species are carnivorous, so feeding would seem to be easy. However, many of the lizards known as "bread-and-butter" species in the hobby are still not kept with the success rates that have been shown by other herp keepers, in particular snake keepers.

This book should enlighten you as to lizard care, especially dietary requirements. It also will give some reasons why many lizard species have had trouble adapting to captive life and, hopefully, will answer some of the questions that are all too frequently asked (usually after a fatality). Keep in mind that nutrition information for many species of lizard is spotty, so keeping those species should be attempted by experienced hobbyists only. The information in this book is meant for the hobbyist intending to keep and/or breed the lizard species that are seen on a common or semi-common basis.

R.D. BARTLETT

The growth in popularity of lizards, such as this veiled chameleon (*Chamaeleo calyptratus*), has brought about a need for accurate information on their maintenance in captivity.

The hobbyist who says a book dedicated to a subject such as lizard nutrition is too specialized probably will not be in the hobby very long. The goal of the successful hobbyist obviously is the proper care of your specimens and an ongoing interest in acquiring as much information on lizard care as possible.

INITIAL CARE

The steps needed before a lizard will stay healthy and breed in captivity can be listed in a pyramid

R.T. ZAPPALORTI

This healthy specimen of leopard lizard, *Gambelia wizlizeni*, has obviously been subject to proper housing and diet, as can be seen by it's robust appearance.

format, where each successive step up the pyramid cannot be reached before the previous step has been satisfied. For those of you familiar with psychology, this is an adaptation of A. H. Maslow's ***hierarchical theory of motivation.*** It looks like this:

breeding
proper diet
proper housing
initially healthy stock

This model can be adapted to any living organism, just as it was changed from the human model to fit lizards. You might notice that some lizards you own are, say, breeding without proper housing, and jumping a step. However, chances are that the minimum basic requirements have been met. This would allow the lizards to advance to the next level, diet (eating), without the apparent need for correct housing. You will find that over time, and your growing familiarization with your animals, this pyramid theory works without fail.

ACQUIRING SPECIMENS

It would seem to some keepers of lizards that buying a lizard would be the easiest thing to do. Just go down to the local pet store and pick what you want. This is not always so. It is vital in the formation of a long-range husbandry program to learn as much as possible of what the prospective species will need for survival in captivity. You also will initially want to know what the particular species looks like in a healthy state. Just letting a salesman pick any specimen from a cage full of animals probably is not wise. Also, picking that thin, "shy" animal in the corner to bring home and rehabilitate is not a great idea either.

The reasoning is that a lizard that looks to be not in the peak of health probably is closer to death than you think. Reptiles, just as many bird species, can hide illness remarkably well. In nature an outwardly sick animal could draw attention from predators, which would increase the chances of the lizard becoming lunch. Whether you are a retailer or a hobbyist, picking appropriate specimens is very important. Here are some points to consider:

1) A tank or cage that has even one animal showing signs of disease (mouth fungus, metabolic bone disease, etc.) is likely to be infested with the disease. Metabolic bone disease is not contagious, but animals in the same quarters are likely to have been subject to the same care and may be close to showing signs of the disease, if not already present. Even if you have been searching for this species for ages or the species is rare or unusual, it would be wise to save your money. Also, normally harmless parasites, such as mites, are a sign that a lizard has been improperly housed and thus has been under extra stress.

2) Look for animals that are full-bodied and active. Bright eyes and good appetites also are ways of determining good health.

3) A problem with many species of lizards today is that they are imported wild-caught specimens. The problem with this is that many

Caution should be taken with wild-caught animals, such as this sand skink (*Scincus scincus*), as they are known to carry both internal and external parasites.

R.D. BARTLETT

R.D. BARTLETT

Above: This *Uromastyx beuti* is a desert species requiring hot, dry conditions in order to thrive. **Below:** Chameleons are not for the beginner. They require specific housing and often hold parasites which need to be removed. Photo of *Chamaeleo labordi*.

R.D. BARTLETT

imported specimens hold internal parasites. In nature this normally would not be a major problem, but under the stresses of capture, importation, cramped housing, etc., parasites can capitalize on the weakened state of a lizard and multiply out of control. Ideally all lizards should be checked by a veterinarian for internal parasites, in particular imported specimens. Some species, such as chameleons, are notorious for this problem. A lizard that loses weight (whether or not it is eating) and shows sluggishness, sunken eyes, or lethargy should be suspected of holding parasites. Such a specimen should not be mixed with others and should be attended to by a vet immediately.

4) Another problem with imported specimens is the chance that they won't adapt to captive life and will slowly waste away and die. Whenever possible buy captive-bred animals to have a better chance of acclimation and to ease the burden on wild populations.

5) Impulse buying can be another problem. The small, cute, colorful Nile Monitor that you see at a pet shop will turn into a 6-feet long, drab, and potentially vicious lizard. It will eventually need a very large enclosure (like its own room) to house it properly.

6) Many species of lizard are known as "difficult" to keep. Some species, such as some chameleons, horned lizards, and *Uromastyx* sp., need specialized diets/habitats to survive. Only experienced keepers should attempt these species.

7) Lizards, like any pet, can get sick. It is important to locate an established veterinarian that specializes in exotic pets, preferably before purchasing any animals. Homemade remedies or those intended for human use will usually hurt, more than help, a lizard, though your pet shop does sell a variety of treatments for common, easily cured ills. Check with local pet stores, public aquaria, and reptile societies for suggested vets.

K.T. NEMURAS

A velociraptor dinosaur? No. A healthy, bright-eyed gecko—*Cyrtodactylus pulchellus.*

R.D. BARTLETT

Top: A hardwood hammock located in southern Florida. Note the lush growths of ferns and bromeliads (*Tillandsia* sp.) that make this area a perfect home for many species of herps. **Bottom:** Improper housing of Green Iguanas (*Iguana iguana*). The cramped conditions are a perfect breeding ground for disease.

W.P. MARA

HOUSING

It will be assumed by yours truly that you have read the above section carefully and understand what was said. It also will be assumed that any and all specimens have been checked out by a veterinarian and quarantined from the rest of your collection.

If these steps are not taken, any guarantee of success cannot be made. Success will then be dependent on the chance that new specimens are free of any parasites and diseases.

Why is this section needed in a book about nutrition? Because a lizard that is not housed properly will not feed, eliminating any need for knowledge of how to feed the lizard.

Initially healthy stock was discussed previously in the section on acquiring specimens. It is the basis for any success with keeping lizards. Once this is satisfied (you have a healthy lizard, free of any parasites, infections, etc.), the next step is reached. Proper housing is the next level and can mean many things, depending on the species of lizard that you are housing. A desert species will need a dry, warm cage, usually with a sand bottom and rockwork throughout the tank. A woodland species will need some warmth, some humidity, branches, etc. An in-depth look into this topic could probably be the basis for

G. DINGERKUS

Above: The low- to medium-sized shrubs and trees make this area of Grand Cayman Island perfect habitat for anoles.

PHOTO COURTESY OF ENERGY SAVERS

Left: Full spectrum incandescent lighting allows captive lizards to produce the vitamin D3 that they require.

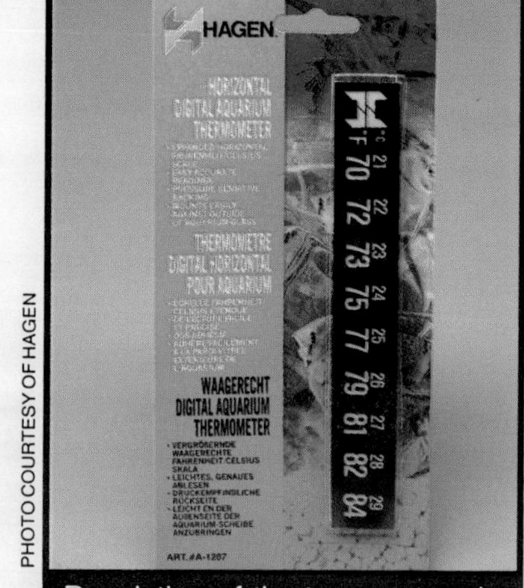

PHOTO COURTESY OF HAGEN

Regulation of the temperature is vital if success is to be had with herps.

another book.

Suffice it to say that housing is possibly the most important aspect of keeping lizards, because it has the most variables involved. Selection of a cage or terrarium is first. Proper substrate and interior decorations are next. Lighting and heat cycles are very important, because lizards are cold-blooded and dependent on artificial lighting to reproduce the seasons (needed in breeding many species). Breeding can be accomplished only if proper companions are housed together (the ratio of males to females).

Lighting for captive animals has come along way since the time when a small incandescent reading bulb was all that was available for both light and temperature control. Photo courtesy of Energy Savers.

Terraria can be easily decorated to appeal to both lizard and keeper. Pictured here is a Bearded Dragon, *Pogona vitticeps*. Photo courtesy of Creative Surprizes.

PHOTO COURTESY OF HAGEN

The smaller plastic tanks that are available can be used as quarantine facilities or permanent housing for smaller species.

It is good to know the origins of the species that you plan to keep. Aquatic and semi-aquatic herps could be found here.

K.H. SWITAK

PHOTO COURTESY OF CREATIVE SURPRIZES

Colorful reptile-oriented decorative backgrounds are attractive and can be useful for shielding the back of the terrarium against distracting motions.

PHOTO COURTESY OF HAGEN

To top a terrarium off, a form-fitting light hood could be added.

Decorative items that simulate natural growths and structures are available at pet shops; these items also can provide refuge.

PHOTO COURTESY OF BLUE RIBBON PET PRODUCTS

NUTRITION

Nutrition is the one aspect of herp keeping that can make or break you in the long run. True, I stated that housing has the most variables, which is correct and which will be the factor that determines success over the short term. However, if you are to have success over the "long haul," then it is imperative that your animals have a proper diet.

NATURAL DIET

The first aspect of nutrition for a lizard (in particular the so-called difficult species) is to determine which niche the species fits into. This does not mean making a field trip to the country that the desired species is native to, but rather investigating through books, pet stores, etc., how the animal fits into its habitat, including what it eats in its natural environment. For most species an in-depth exploration is not necessary, as the information that is available in books is usually more than enough to set up a complete nutrition program. The niche of an animal can be determined by three factors: its natural climate, habitat, and the predator/prey relationship it has with other species living in the same area. All this means is whether it is warm or cold where the animal lives (climate), the vegetation (woodland, desert, swamp, etc.) that inhabits this area (habitat), and the animals that the lizard eats and is eaten by

A lizard such as a Green or Plumed Basilisk (*Basiliscus plumifrons*) is a rainforest species normally living near bodies of water.

R.D. BARTLETT

The Blue Rock Lizard (*Petrosaurus thalissinus*) is a desert species native to the Baja California region.

(predator/prey relationship).

The natural diet of a lizard is the key to a captive diet for a lizard. With this in mind, and an idea of the lizard's niche, it can be possible to duplicate or at least simulate the animal's diet. Some lizard species will require duplication of their natural diet to thrive in captivity, but most species will prosper (and breed) on a diet *simulating* their natural diet. A simulated natural diet is one in which prey items that don't occur in an animal's natural range are substituted for animals that normally would. For instance, a Common Agama (*Agama agama*) might eat locusts in its native Africa, but usually will take crickets as a substitute with little or no ill effects. Locusts are closely related to crickets and are nutritionally very similar. A species that has been quarantined and housed properly and is not eating a diet of simulated natural foods obviously will need to have its diet altered to allow some or all natural food items to enter the diet. An animal that currently is on a diet composed of natural foods often can be gradually changed over to a simulated-type menu, as long as the simulated menu is comparable nutritionally to the natural diet.

Usually with some basic investigation the natural foodstuffs can be ascertained and plans can be made to set up the diet. A simple example can be shown in the common Green Anole (*Anolis carolinensis*). First a natural geographical range is identified. This would be the southeastern United States. There it is native to woodlands and scrublands where there is plenty of cover and not too much sunlight (usually in the form of filtered sunlight). The air has some amount of humidity, anywhere from the drier air of scrublands in southern Florida to the very humid inland portions of the Carolinas. The anole forms territories in bushes, shrubs, and smaller trees. The insect fauna that inhabits these areas would normally include crickets, grasshoppers, flies, mosquitoes, ants, and beetles, as well as a variety of spiders and other small creepy things. This is what a wild Green Anole would eat (plus or minus some of the items listed here, and with the possibility of many other species of invertebrates and baby anoles included in the menu, too).

So now we have an idea of what this lizard eats in the wild, but also of major importance is what the insects have been eating. This would include both plant and animal matter, and this is what is being taken in by the lizards, via the insects. Thus a nutritionally complete diet is attained.

FEEDING IN CAPTIVITY

There will need to be some compromise in captivity as far as the actual selection of insect species is concerned, because it is unrealistic to consider every natural species of prey that the lizard would eat as attainable. True, one could go out and field collect many varieties of insects and spiders for a lizard, but even this would not have everything that the lizard's natural diet included. If the species of lizard is a "difficult" type, then some of its natural diet may have to be obtained, either

An Alligator Lizard (*Gerrhonotus multicarinatus webbi*) is a woodland species native to the southern U.S.

R.D. BARTLETT

through field collecting them or getting them through commercially raised stock.

For virtually every herp keeper at least some reliance on commercially raised insects will be needed. From the occasional cricket as a *treat*, to every meal that the lizard will ever have, some reliance on a good pet store will need to be relied upon.

Once you have an idea of what your species of lizard eats in its natural habitat, it is just a matter of adapting a simulated natural diet for that lizard species.

COLLECTING INSECTS

This section is for those of you who are "serious" herpetoculturists and are willing to go to any lengths to see to the well-being of your collection. For the person living in the city, this section may be a waste of your time unless you are willing to travel to the remote, uncharted wilds of the suburbs. For those living in colder climates, you are basically forced to be a part-time collector of bugs, as the winter is a "dead" period for collecting. The added nutritional content of wild-caught insects cannot be equaled with captive-raised fare, no matter how well they are loaded. Lizards subjected to the same diet day in and day out tend to lose some of their natural appetite, which might make them a bit more susceptible to illness. (Actually, this is highly arguable at best, so take it with a grain of salt.)

The anoles are a family of woodland species that specialize at eating insects. Pictured here is the Jamaican Giant Anole (*Anolis garmani*).

P. FREED

R.D. BARTLETT

A Sungazer, *Cordylus cataphractus*, is a species that would benefit from wild-caught insects.

For those who live near or are willing to travel to areas that hold large numbers and varieties of wild insects, collecting is for you. For small to medium sized collections of insect-eating lizards, all that is needed is several minutes in a field or pasture to secure a week's worth of insects. A large collection of lizards would only require a few more minutes and you would be "crawling" with them.

The most common method of collecting insects is the "sweeping." The name is self-explanatory. A net similar to a reinforced butterfly net is swept or dragged through a field. At frequent intervals the net is checked for any insects that have fallen into it and the insects are placed into a jar or fine mesh cage that won't allow escapes. The net can also be swept through bushes, small trees, or anything that sports a lot of vegetation.

Another method is to search out the hiding spots of the insects. The usual homes of insects are under logs, near any body of water, etc. This is a good way of capturing insects when a field is not a practical place to sweep.

A third method of capturing insects is the use of some sort of light and a sheet. The "night hunt" (also known as light collecting or sheeting) is a very easy way of attaining moths, beetles, and any other nocturnal, flying bugs. A light (propane lanterns and ultraviolet lights are especially effective) is placed in front of a white sheet or piece of plastic on a moonless night with little wind. If you can place the sheet near a stream or lake, so much the better. Within minutes the sheet will be bombarded with various insect species, and it is just a matter of collecting them for your own use. This method can be used in a back yard just as easily as a field and is probably the best choice for someone who can't get out to the field to collect insects but still wants to give his lizards the best.

THE CHOICES

The diet of your lizards usually will be made up of commercially raised insects. The selection that is available to the reptile keeper has increased steadily in the last few years as interest in herps has increased. So what is available and what is desirable to the hobbyist through pet stores, live food dealers advertising in reptile and fish magazines, and biological supply houses? Here's the list of the species that can be relatively easily attained and information on the species that can be bred practically in captivity.

M. GILROY

The intended goal of this cricket should be food, not friend.

CRICKETS

Crickets usually are the most commonly available insects encountered in the pet trade and can easily be purchased almost anywhere reptiles are sold. There are good reasons for this. Crickets are the insects in most demand by hobbyists because the cricket is a natural food source for many species of lizard. For lizards that don't include this species on their list of natural prey items, they are extremely nutritious insects and most lizards are all too happy to adjust and include them on their diets. Additionally, crickets are very easy to keep and culture. Crickets will eat just about anything organic (and sometimes inorganic, such as synthetic carpeting) and, just as long as they are kept warm and fairly dry, they will do fine.

There are many species of crickets known to science, but there are only two types that come to mind when a herp keeper thinks of crickets. One type is the common field crickets, *Gryllus* of various species, that usually are shiny black in color. This is the type that is common over much of the United States and Europe in the summer months, chirping away in the warm nights. Field crickets should be used by the herp keeper with caution. The black shininess of the chitinous "shell" gives some indication of the thickness of the cricket's armor. The field crickets generally are somewhat larger and bulkier than the gray cricket and should be avoided when feeding smaller lizards. The thick skin of field crickets is tough to digest. Normally only larger specimens are captured, and they can cause all sorts of digestive problems for a small lizard that manages to swallow one. Additionally, if a lizard does not immediately eat this cricket, come nightfall when the lizard is asleep and the cricket is foraging for food, the cricket may do damage to the lizard itself.

However, for larger species of insect-eating lizards such as collared lizards (*Crotaphytus* spp.), basilisks (*Basiliscus* spp.), Tokay Geckos (*Gekko gecko*), and small monitors, field crickets can make an excellent addition to the diet, especially if gut-loaded or coated with vitamin and mineral supplements.

The other cricket commonly encountered by the herp keeper is the Gray Cricket, *Acheta domesticus*. This is the species that should, for many herp keepers, be the most easily identified and used species. It likes warmth and does not do well outdoors in cool climates, though it originated in southern Europe. This is the species farmed by breeders in the southern U.S. It is a soft-bodied insect, as compared to the field crickets, and thus is suitable for virtually all insect-eating lizards. It is sold in many sizes, from pinheads (recently hatched) to fully grown inch-long adults. The low ratio of indigestible chitinous shell to edible internal mass makes this a nutritious and efficient basis for many diets. This species is easily raised, gut-loaded, and vitamin

The Gray Cricket is the main component in the diet of most captive insectivorous lizards. Photo: W.P. Mara.

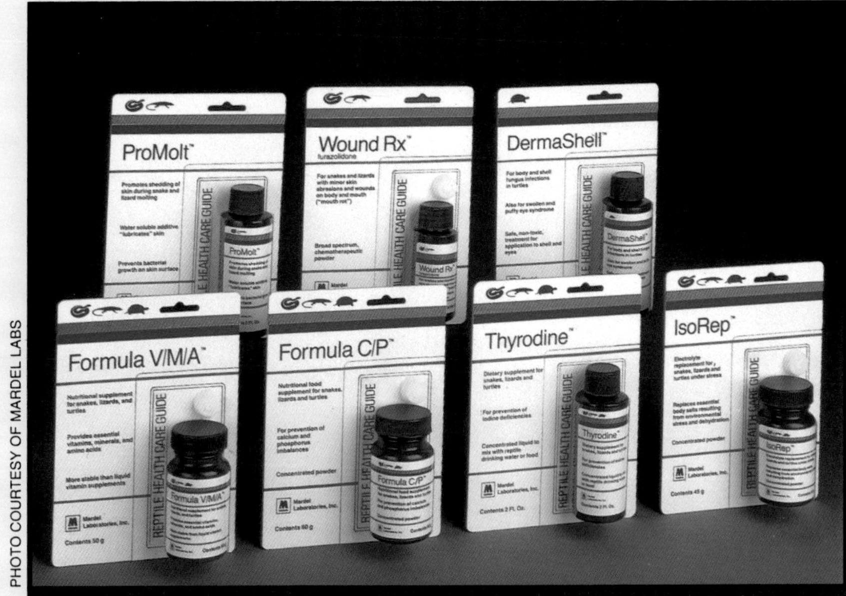

PHOTO COURTESY OF MARDEL LABS

The rise in popularity of herps in recent years has brought with it many new medications and vitamin supplements, specifically made for reptiles and amphibians.

The House or Field Cricket (*Gryllus* sp.) can be used for feeding larger species of lizard, such as leopard lizards and agamas.

M. GILROY

dusted for offering to lizards. It *definitely* should be included in the diets of lizards, if the lizard is willing to eat it.

Breeding crickets is fairly simple to accomplish. This allows you to have the several sizes of crickets that may be needed in order to feed different sizes of lizard. First you will need a container to house the cricket colony. Standard sized aquariums make the cheapest and most efficient method for housing the colony. A 10- or 20-gallon tank is perfect. Wooden containers can be used, but they are difficult to clean and probably are more costly to buy or make. The tank should be cleaned thoroughly. A layer of soil (one-third fine loam, one-third peat, and one-third fine sand or equivalent) approximately 2 inches deep should be laid down. This soil will need to be kept moist *but not wet.* A wet substrate will lead to a smelly, spoiled culture of dead crickets. A low relative humidity is what you are looking for.

PHOTO COURTESY OF FOUR PAWS

A clean, efficient way of maintaining a reptile's tank is with the use of pine bedding.

With the substrate now in place you can add some shelter for the crickets. Egg cartons make an excellent choice. They are cheap, will last a long time if kept dry, and have a lot of surface area to house many crickets in a relatively small area. The egg cartons can be stacked to several inches from the top of the aquarium. A tight-fitting cover is your next requirement. Crickets will jump to the cover, and if the container is not tightly covered you will have more crickets in your house than in your tank!

Foods for the crickets can be oats, corn meal, pieces of fruits and vegetables, or one of the commercial cricket diets now available. These foods can be rotated for best effect. Water will be acquired from the fruit. The food should be placed in shallow dishes where it will be easily available to the insects and easily taken out for cleaning. The "wet" foods (fruits and veggies) should not be kept in the tank for more than two days. Dry foods can safely be left in the tank for a week or so with no ill effects observed.

Warmth is essential in raising these insects. An incandescent light mounted above the insects housing will do nicely. An ambient air temperature of 80-82°F is suitable. Undertank heaters can also be employed but are initially more costly to purchase. Once the temperature, shelter, hiding spots, and food are in the tank, a group of crickets can be

Grasshoppers can be a valuable addition to the diet of captive lizards.

G. DINGERKUS

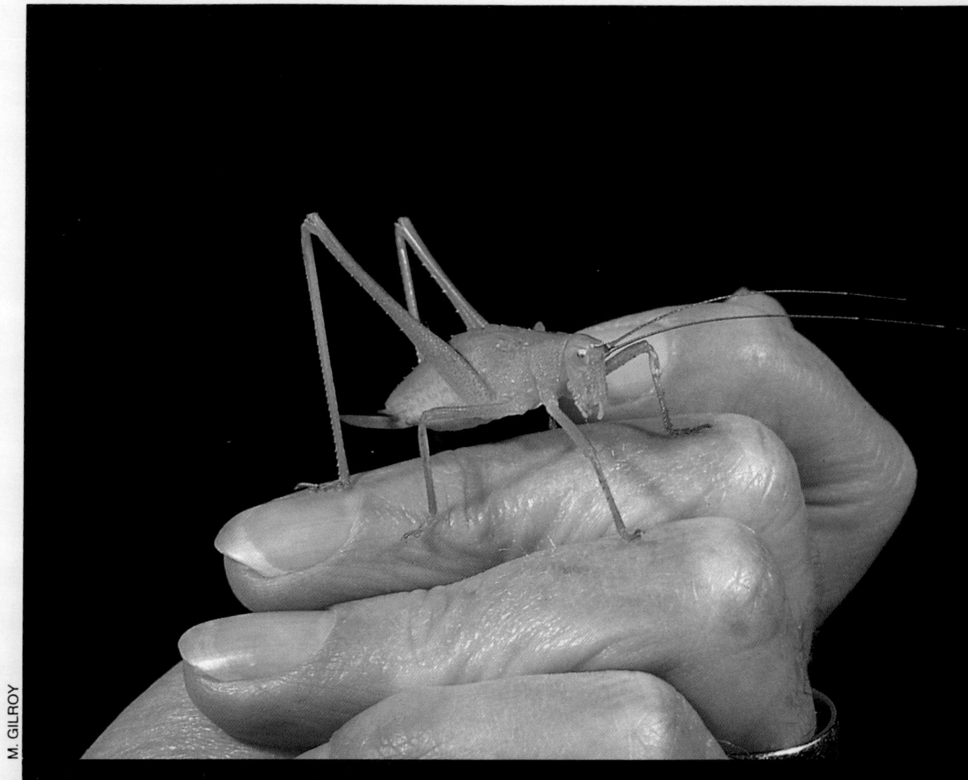

M. GILROY

Above: Katydids occur throughout many regions and are a natural prey item for many species of lizard. **Below:** This species of grasshopper is native to Belize, where it might be a regular on the diet of basilisks and ameivas.

P. FREED

G. DINGERKUS
The Lubber Grasshopper, pictured here in the Florida Everglades.

introduced to start the colony. You will find that in a few weeks you have crickets of many sizes constantly at your disposal.

GRASSHOPPERS

Since we are on the subject of crickets, grasshoppers should be mentioned. Grasshoppers are a close relative of the crickets and thus are anatomically and nutritionally very similar. They eat mainly vegetable matter but will scavenge dead animal protein as well. The genera that are used most often by herp keepers include *Melanoplus*, *Romalea*, and *Dactylotum*, but any smaller grasshoppers swept from clean (insecticide-free) vegetation will be eaten by lizards of appropriate size.

The grasshoppers seldom are bred by commercial breeders (though a few large locusts are available on the European market) and are best obtained by catching them yourself. They do make an excellent addition to the diet of larger lizards, and you will find that your lizards are all too happy to make them a part of their diet.

MEALWORMS

This is the insect species, besides the cricket, that is carried most by pet shops. There are several species that are known, but the one that the pet stores carry is *Tenebrio molitor.* Known also by the names darkling ground beetle, tenebrionid beetle, and pincate beetle, it generally is referred to as the mealworm, and use of the other names will usually get you some strange looks. These insects have become a controversial point among herp keepers lately. Some keepers say that mealworms are too chitinous and nutritionally deficient to be considered as practical for a lizard's diet. They have also been known to cause digestive problems, including regurgitation in animals allowed to gorge themselves on them. These points will all be true if a lackadaisical approach is taken to food preparation. In other words, if you buy a portion of mealworms, bring them home, put them in a dish or bowl, and then put them directly in the lizard's terrarium, this is exactly what will happen. To make mealworms worth using, some preparation is needed.

There are several other species of "mealworm" that can be cultured for use by the herp keeper. *Blapstinus moestus* is very similar to *T. molitor. Alphitobius diaperinus*, known as the lesser mealworm, is a quarter-inch mealworm, very similar in appearance to *T. molitor.* They are

Several species of lizards make ants an integral part of their diets. Pictured here are leaf-cutter ants from Guyana.

P. FREED

C.O. MASTERS

sometimes found in musty old carpeting and usually can be cultured similarly to the other mealworms. *A. diaperinus* can make a good addition to the diet of smaller lizards, such as Green Anoles and geckos.

The other most common mealworm encountered is the king mealworm, *Zophobas* sp. The growing popularity of using this insect as an integral part of medium to larger sized, insect-eating lizards' diets is due to two advantages. First, they are known to have a smaller chitin-to-body mass proportion than *T. molitor.* This means easier digestion of even the full-grown harder-bodied adults. Secondly, raising them yourself means that a larger variety of sizes will be available (simply because they get larger than *T. molitor*) to you to be able to correctly size the lizard's food. The down-side to this species is that it is native to warmer climates and so must be kept warmer than *T. molitor* in order to thrive. They will be sluggish at room temperature and will need to be kept warmer if

Above: An old stand-by, the Mealworm is commonly carried by pet stores, and can make a valued part of a lizards diet. **Below:** The Giant or King Mealworm has been gaining in popularity as a food source for larger lizards.

M. GILROY

breeding them is your goal.

Raising these insects yourself will allow proper nutrient loading and size selection, thus maximizing the value of this all too mis-used animal.

A large breeding colony is the best way to go about using these insects. This will allow mealworms of many sizes to be available at any time (this includes the adult beetle stage). Also, freshly molted mealworms will be available all of the time. These mealworms are white in color (their exoskeletons have not yet hardened) and are much easier to digest by lizards.

This is one of the easier species of insect to breed. They can be raised in a fish tank the size of a 2.5- to 5.5-gallon aquarium (12 to 16 inches long by 8 to 10 inches high by 8 to 10 inches deep). The aquarium is set up in a layered fashion. A half-inch layer of oats or alfalfa meal is laid down on the bottom of the tank. This layer is covered with burlap or cotton cloth. You repeat this until you have several layers that are a few inches thick. Pieces of fruit or vegetable are placed between the layers (one per layer) and changed every two to three days. The tank should be covered to prevent the adult beetles from possibly flying out. Now a portion of mealworms can be added. The colony should have new oats or alfalfa meal added as necessary to keep a steady supply of food ready for the mealworms. These insects do best in some warmth, and an incandescent bulb overhead should do the trick. The king mealworms will need a higher temperature to do well (same as Gray Crickets).

The beetle of the Giant Mealworm is eaten by only the most aggressive of eaters—collared lizards, agamas, etc. Photo: M. Gilroy.

FRUITFLIES

This is the species (*Drosophila melanogaster*) that has been a favorite of scientists for so many years. It has an extremely fast egg-to-egg generation time (the time it takes an egg to hatch into a larva that matures to adulthood and finally lays its own eggs), which is usually 10 to 14 days. It has few pairs of chromosomes (six), and the fly can be bred for mutations with relative ease. This has resulted in literally dozens of new, fixed strains, including the wingless variety that is of importance to herp keepers. Without the wingless trait fruitflies would be virtually unmanageable. But because this strain can't fly it can be raised and bred easily in small containers.

The fruitfly makes an excellent food for small lizards, including small anoles (*Anolis* spp.), geckos, skinks (*Eumeces* spp.), and fence lizards (*Sceloporus* spp.). They also are excellent first food for babies of many species that eventually grow to larger sizes. Once again the proper nutrient loading of these insects must be stressed.

Fruitflies are easy to culture, especially with the products that many biological supply companies offer. Culturing mediums, breeding vials, anesthetic, etc., are all easily acquired and make culturing these flies easy. However, you can breed them without these products. Mason or pickle jars can be used with a culturing medium of mashed fruits and a bit of bran mixed in. Several tablespoons of this mixture are added to the bottom of the jar. Resting places in the form of broom straws or pieces of cardboard are added. Then a stock of fruitflies (preferably the wingless type) can be added. A piece of cloth is fastened around the top with a rubber band to prevent escape, yet allow ventilation through the jar. Separate colonies should be set up several days before use of the culture for feedings. This will allow you to have flies ready all the time and allow you to clean the jars, as they readily foul.

HOUSEFLIES

The common housefly, *Musca domestica*, can make a good addition to the diet of lizards that are too big to pick at fruitflies. Because they are so closely related to the fruitfly, they can be considered nutritionally very similar as well. This is a species that can only be captured outdoors, as they are not cultured like their cousin the fruitfly. However, they are easy to collect from the outdoors. All that is needed is a plastic soda bottle with a piece of fruit placed inside. Leave the bottle outside for a day or so, and you should have a supply of flies, possibly along with other species of insect.

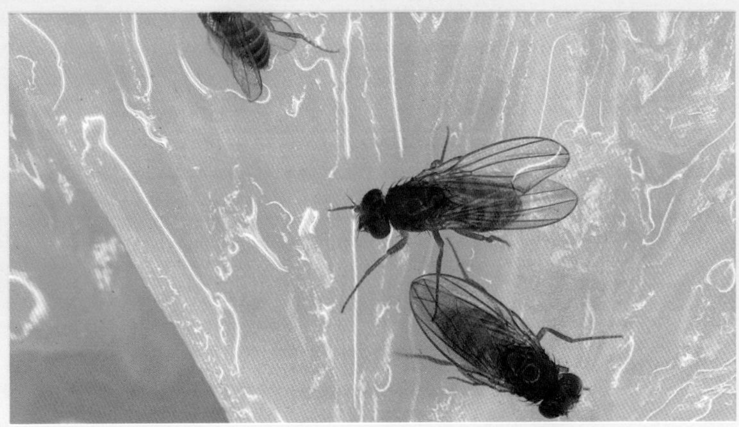

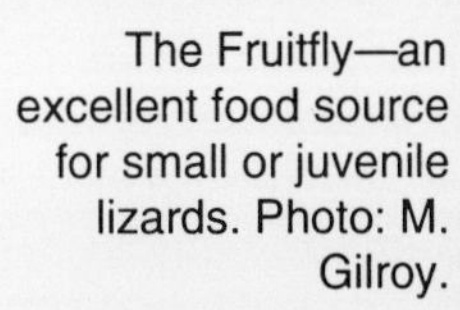

The Fruitfly—an excellent food source for small or juvenile lizards. Photo: M. Gilroy.

COCKROACHES

These insects make even the experienced herp keeper cringe, not necessarily because the insect is "creepy," but because of their possible escape and subsequent colonization of one's house (and the need for an exterminator). This is a distinct possibility with some of the pest cockroach species, such as the American Cockroach (*Periplaneta americana*), the German Cockroach (*Blatella germanica*), and the Oriental Cockroach (*Blatta orientalis*).

There are several species that are bred by biological supply companies that are suitable for feeding herps.

This includes *Blaberus craniifer.* Captive-bred strains of this cockroach include wingless types and types that cannot climb glass because of atrophied leg hairs. These are the types to use for culture for lizards. They can be raised with relative ease and are a nutritious meal for any medium to large-sized lizard. It is *highly* recommended that you not try to raise domestic pest cockroaches, as they are excellent escape artists and might just make themselves at home in your house.

You will find, possibly to your surprise, that many lizards love cockroaches. Lizards such as collared lizards, Leopard Lizards (*Gambelia wislizenii*), and smaller monitors (*Varanus* spp.), as well as a host of other species, will devour these "pests" with gusto.

Cockroaches can be set up to breed in colonies exactly like the cricket setup described above. Use of the cockroach mutations that cannot climb glass is highly desirable.

ANTS

Ants can be an excellent food source for many small lizard species. This includes horned lizards (*Phrynosoma* spp.), earless lizards (*Holbrookia* spp.), and flying dragons (*Draco* spp.). Usually ants can be found all around buildings, and occasionally in them! They can be harvested easily by leaving a piece of fruit or even a potato chip near the entrance to an ant hill. In an hour the bait will be covered with ants, which can then be harvested and fed directly to your lizards. Do not overfeed, as they easily could make a pest of themselves once inside your house.

WAXWORMS

The waxworms, *Galleria* and *Achroea*, are the maggots of the wax moths. Waxworms can be obtained through fishing tackle stores as well as many pet shops. The commercial waxworm is very difficult to raise, because it needs bee honeycombs in order to survive. The waxworm can make a very good supplement to the diet of many species of lizard, though. Waxworms are sold packed in wood shavings and should be kept refrigerated in order to slow down their metabolism, otherwise they would starve to death. Obviously this is a species that cannot be nutrient-loaded, but it can be vitamin-dusted like most other insect species. They can be taken out of the refrigerator as needed to feed to your lizards. The adult moths of this species can be readily fed to your captives, and they will be eaten hungrily, too.

It should be noted that, although many lizard species devour these "worms" readily, they should not make up a major portion of the diets of these lizards. Waxworms tend to be very fatty and are not the easiest of insects to digest. If fed in excess they can cause obesity; if feedings of these continue they can cause intestinal blockages. This should not cause you to overlook this potentially positive addition to a lizard's diet, though.

SPIDERS

If you have ever fed a spider to one of your lizards you will know that lizards love them! Along with their distant relatives the "daddy longlegs" (harvestmen), spiders can make a great addition to the diet of lizards. They are not culturable to the extent needed for feeding lizards, but can be captured from the out-of-doors. There are many species that can be used, but make sure that they are small enough for your lizards to easily overpower them, as the larger species can deliver a nasty bite. Fed correctly, though, they can round out the diet of a lizard quite nicely.

G. DINGERKUS

Butterflies, when captured in areas that are free of pesticide and fungicide use, are an excellent treat for lizards.

MOTHS AND BUTTERFLIES

This is another group of insects that really cannot be cultured indoors but nonetheless are an excellent addition to the diets of captive lizards. You can find them wherever flowers grow, often in the company of bees. They are easily harvested with a small mesh net and can be fed directly to your lizards. It must be mentioned that insects should not be harvested in areas that have been sprayed with any man-made chemicals, i.e., pesticides, fertilizers, fungicides, etc. These substances can be extremely toxic to your lizards and can kill them with little or no warning. Use your own judgment as to where it is safe to hunt for insects.

Spiders come in many forms and sizes are relished by lizards. **Top photo:** *Nephina clavipes*, by S.A. Minton. **Bottom:** *Loxosceles laeta*, by A. Perucca.

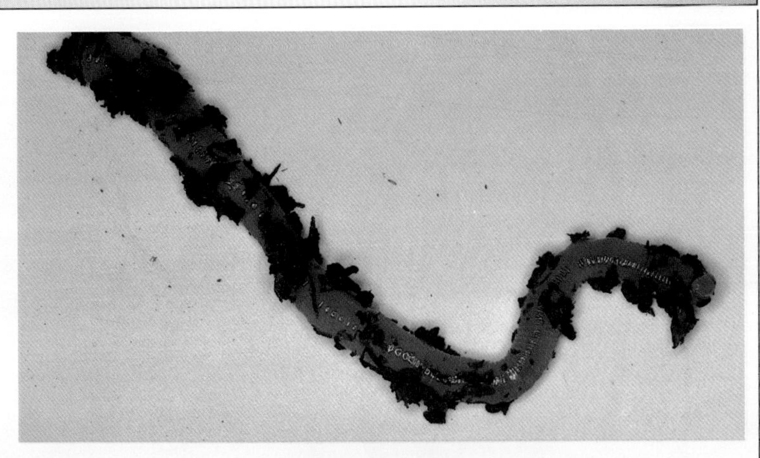

Various species of agamids make earthworms a part of their usually insectivorous diet. Photo: M. Gilroy.

EARTHWORMS

True, earthworms are not insects, but they are eaten by some species of lizards that are known as "insect-eating." A highly nutritious supplement to those species that will take them (most species will not), they are easily found in any area with moist soil and also can be purchased in any bait and tackle store, as well as through many pet stores. Some types of lizards that may take these annelids are mountain dragons (*Acanthosaura* spp.), tree dragons (*Gonocephalus* spp.), and sheltopusiks (*Ophisaurus apodus*).

Earthworms can be set up to breed easily. Wooden or plastic trays (20 x 20 x 6 inches high) can be used. Aquariums can also be used. The tray or aquarium is filled approximately 5 to 6 inches deep with a good quality soil. Some oatmeal or other grain product is sprinkled over top of the mixture. Several small pieces of potato are added in to the soil. A piece of burlap should be added to cover the culture, allowing some

Moths are a nutritious way of varying the diet of captive lizards. Pictured—Cecropia Moth (*Hyalophora cecropia*).

W.B. ALLEN, JR.

moisture to remain. Water should be added as necessary to keep the culture moist, but not wet. Earthworms can now be added. This should give you an ample supply of earthworms throughout the year. (In such a small setup mold may be a major problem, however, and it is easy for the culture to overheat or to become too wet or dry. Additionally, most earthworm species rapidly will ammoniate the soil in a container where the bottom does not have air access, leading to rapid colony die-offs. Frankly, unless you need large numbers of earthworms, it is easier in most cases to collect them or buy them.)

PINKIES

A pinkie is a newborn mouse. It is obviously a vertebrate and normally would not be on the diet of an *insect-eating* lizard, but many of the larger species of lizard are opportunistic and will consume pinkies quite readily. Pinkies are an excellent food for lizards. They have a high nutritional value, and combined with a base of insects make quite a menu for larger lizards.

They should not be forced on lizards that are too small to take them, as it would be cruel to both the lizard and the pinkie. Tokay Geckos, collared lizards, and basilisks are just a few of the many species that will accept pinkies as part of their diets.

The breeding of these rodents is a fairly simple procedure to accomplish. A standard 10-gallon tank is perfect. Put a 1- to 2-inch layer of cedar or pine chips or corn cob bedding on the bottom of the tank. This substrate should be

A Leopard Gecko consuming a pinkie. This normally insectivorous lizard is an opportunist and will readily take these as food.

J. MERLI

R.S. SIMMONS

A Glass Lizard (*Ophisaurus compressus*) will take a wide range of invertebrate food.

changed weekly. A water bottle should be hung from the side of the tank and the water in the bottle should be changed every day or two. A vitamin solution can be added to the water as per instructions that accompany the vitamins, usually two to three times per week. The mice generally are more productive with these vitamins added. The water bottle should be thoroughly washed out weekly.

A tight-fitting cover should be purchased or made that will not allow the mice to push or chew their way out. A hiding place in the form of a ceramic or wooden ornament or box can be used to give the mice refuge. A ceramic or plastic bowl for food should be provided. This bowl should be filled with a commercial rodent seed mix every day. Now add a group of mice to the tank. A group of three to five females can be added with *ONE* male (more than one male per tank might lead to fighting).

This is the ideal setup for mice if you are looking for a maximum number of babies. Once this breeding colony is established, it should be about three to four weeks before babies (pinkies) start to be produced. Pinkies that are not used for food obviously will grow to adulthood in the tank. Ideally your group of breeding adults should stay at the same number of animals. A crowded tank seems to shut down the mice's instinct to breed. This setup should give you a constant supply of pinkies that can be used for feeding.

There are many other species of insects and invertebrates available that are beyond the scope of this book to include. You should be able to use the insect species discussed here to formulate a suitable diet for any but the most specialized of lizard species. Insectivorous lizards will not need other insect species to thrive, but other insect species can be included to help round out their diet.

DIET SELECTION

You now should have a relatively good idea of the insects that are available for feeding to your lizards. With this knowledge it should be possible to figure out what species, as well as what *size*, insects you should feed to your lizards. Those herp keepers who give attention to insect selection often overlook the idea of correctly sized food items. It is very important, especially in delicate species, that food items be of the correct size for feeding in order to allow the lizard to digest the insect properly.

As an example, it is a very good idea to feed crickets, mealworms, and cockroaches to House Geckos (*Hemidactylus frenatus*). However, a house gecko grows to only 4 or 5 inches in length. If adults of these insect species are fed to this gecko, the lizard will suffer. An adult cricket, mealworm, or cockroach can be an inch or more in total length. This is approximately one-fourth the body length of the lizard; the gecko will have a very difficult time digesting such a large meal. In fact, it often will regurgitate a meal this large a day or so after it has eaten (often without the owner's knowledge; the owner will assume that the lizard has consumed and is digesting the meal without difficulty).

B. KAHL

A Snail Skink (*Tiliqua gerardii*) eating from its owner's hand.

So how can you guard against this sort of thing happening? Correctly size the insect prey and the problem is solved. In nature small lizards such as geckos and anoles normally prey on animals smaller than you would imagine. The usual fare for these wild lizards is small spiders and insects that are approximately one-quarter inch to one-half inch in length. The lizards constantly are hunting for these insects throughout the day. This allows a wild gecko to eat 10 to 20 small insects in the course of one day's foraging. In captivity, if you were to feed a gecko one "giant" cricket per day, it should theoretically provide the nutrition of many smaller insects, but because the lizard may regurgitate the insect or at best only partially digest the insect, it really makes sense to feed smaller fare.

Facing page: A tame Red-sided Skink, *Mabuya perroteti*, makes short work of a cricket. Photo: B. Kahl.

Use common sense when feeding your captives. A lizard should not take 10 or 15 minutes to swallow a meal. Normally it will take a lizard under two minutes to completely swallow a correctly sized prey item.

Lizards are not snakes! Unlike many snakes, which hunt for food once a week or so (and have a comparatively slow metabolism), lizards generally are more active than snakes and eat on a much more frequent basis. It is of great value to try and simulate the lizard's natural habits by feeding smaller insects on a more frequent timetable.

Here are the three keys to proper selection of food items for an insect-eating lizard:

1. SPECIES: Choose the proper species of insect to feed to the species of lizard in question.

2. SIZE: Use your best judgment in picking the correct size of insect that will give your lizard the most nutrition.

3. NUTRITIONAL VALUE: Determine the times when these insects will be most nutritional (often directly after molting).

W.B. ALLEN, JR.

These Gray Crickets are being fed an orange with added vitamin/calcium powder to increase their nutritional value.

CONDITIONING INSECTS

Now that you know what is needed to correctly select and size insects for your lizard, you can proceed to the next step, conditioning your insects.

As a general rule, when it comes to buying insects from a pet store the insects that you purchase will be essentially void of most of their possible nutritional value. These insects probably have not eaten in days, often weeks! This means that they are "skin and bones" and offer a lizard not much except for something to crunch on. The idea is that whatever an insect eats will be exactly what the lizard eats. *You are what you eat.* It would be safe to assume that the wholesaler and retailer of these insects have not properly fed them, so it is up to you to correct this. If you are culturing your own insects they should be nutritionally "loaded" for feeding.

Just a little work to prepare these insects for your lizards will go a long way in making you a success in keeping and/or breeding your captives. Follow this simple plan to guarantee that the insects you feed your lizard will be "gut-loaded" and full of nutrients.

take them home and house them in a tank or cage from which they cannot escape. For winged insects this means a container with a cover. Also make sure that species with cannibalistic tendencies, such as crickets and praying mantises, have ample hiding spaces. Egg crates and paper towel tubing are good for this purpose.

Once properly housed they can be fed a proper meal. Many insects are opportunistic and will take similar food items. Many different foods can be offered. Rotate these foods to give the lizard the best nutrition possible. Thin layers of oatmeal, corn meal, and bread crumbs can be one possibility. Pulverized dog or cat food spread along the bottom of the cage is another idea. Tropical fish food, be it the flaked or pelleted forms, is another choice. Lightly sprinkle the food with a calcium/phosphorus

J. DOMMENS

A Knight Anole (*Anolis equestris*) devouring an unfortunate grasshopper.

You can rule out feeding insects to your lizards on the same day that you purchased the insects. These insects will need a full day of feeding in order to be considered "nutrient-filled." Wild-caught insects can and should be fed to your lizards on the same day as they are caught. They should have been eating all sorts of vegetation that is nutritious for your lizards. Nutrient-loading these insects is redundant.

Upon purchasing your insects,

powder (often sold at pet stores). Switch these three groups regularly for the best effect.

Food insects will need a source of water. A water dish in with the insects is not the best of ideas, as the insects usually will crawl into the dish and quickly drown. A better idea is the use of pieces of fruit or vegetable placed in the insect container. Slice up a piece of citrus (such as orange, lime, or grapefruit), apple, potato, carrot, cantaloupe, etc. These, too, should be varied on a regular basis. Just a slice placed on top of the insect's food will do the trick.

The insects should be allowed to eat these foods for approximately one full day before being fed to the lizards. Any longer and you might experience losses from your insect stock. Any shorter and the insects may not have been given enough time to fill themselves. Once nutrient-loaded, they can proceed to the next step, and that is vitamin and mineral dusting. Don't reuse the food items of the insects; they foul after a day or two of use.

It seems as though a lot of work is needed just to keep a little anole alive, but you will find that your work will become much easier once you know exactly what you have to do. You will find that there will be a major difference in your lizard's health when a proper diet is given it, as opposed to just throwing some crickets in the cage once a week.

DUSTING INSECTS

We have now picked the correct insects for use and conditioned them properly. We must now "dust" the insects with a vitamin and mineral powder supplement to round out the nutrient requirements of our insectivorous lizards.

A question you might ask is: "If we have bothered to condition the insects prior to feeding, why should we have to use a vitamin/mineral supplement?" The answer is that no

A mealworm culture. These insects are one of the easier species to culture, as they will thrive in small aquariums or plastic shoe boxes filled with oatmeal.

W.P. MARA

M. GILROY

Old World lizards, such as agamas, naturally feed on Old World insects, such as this African Locust (*Locusta migratoria*).

matter how well you condition your insects, they are still missing or deficient in some vitamins and/or minerals necessary for the long-term well-being of your captives.

Dusting is possibly the easiest of the steps to take. First, however, you should know something of what vitamin/mineral supplementation will do for your lizards. It will accomplish exactly what vitamins and minerals will do for humans, that is, ensure that any vitamins or minerals absent or deficient in the diet will be supplied. Currently, information on exact nutritional needs of insectivorous lizards is lacking, so we are left with our instincts and knowledge of the natural histories of these lizards to figure out what kinds and how much supplementation should be used.

There are two types of supplements that ideally should be used. The first type is a vitamin/mineral mix that is produced by many manufacturers under various titles. Consult the side panels of the containers for the ingredients and nutrition information on the contents. Compare and contrast the types available to you for the one that contains the widest variety and the best quality ingredients. Cost should not be a factor.

The second type of supplement that should be used is a calcium/ phosphorus mix, also sold by various manufacturers under numerous titles. Most of these calcium/

phosphorus mixtures are formulated to allow maximum uptake of these minerals by reptiles. A ratio of 1.0-1.5:1.0 calcium:phosphorus is a good balance. Asking your veterinarian or an experienced pet shop salesperson (who specializes in reptiles) about which particular supplements to purchase might be the best thing you can do to ensure that you have the correct supplements.

It is also an excellent idea to inquire as to how to use the supplements correctly. Most manufacturers of these supplements will tell you to use them on a daily basis, but most veterinarians and pet shop salespeople will tell you this is too much. I would agree with the veterinarians on this one.

Here are the general rules that go along with supplementing a lizard's diet:

1. The larger the lizard, the less frequently it will need supplements. This means that larger species need less supplementing than do smaller ones. It also means adults of a species will need less supplementing than juveniles of the same species. The reasoning is that larger lizards normally will eat less frequently than smaller ones, but their meals are larger. Their metabolism is just that much slower.

A juvenile of any species obviously is growing and needs more supplementation to fulfill the needs of its growing body than an adult of the same species.

2. Vitamin/mineral supplements *can* be used on insects that have not been conditioned beforehand. It is much more desirable, though, if you first condition your insects.

3. Consider the size and type of species that you have and get yourself on a regular schedule of when to supplement. If you are anything like me, if you don't make an effort to get into the habit of doing something important, you will end up not doing it correctly.

4. A good blend of supplements is two parts of the vitamin/mineral mixture to one part of the calcium/phosphorus mix.

The actual method of vitamin/mineral supplementation is as follows. Many adjustments can be made and there are no rules carved in stone.

1. Apply the vitamin/mineral and calcium blend to the insects by putting a small portion of the blend (approximately one-half teaspoon) into a plastic food storage bag. Drop in the insects that you want to dust. Close the bag and shake lightly, allowing the powder to evenly coat the insects. The insects can then be fed directly to the lizards.

2. Crickets tend to hold the blended powder the best. Other species can be used, but refer to the next step to find out when to use the supplements.

3. Baby lizards usually will need to be fed more than once a day. They often have very short digestive tracts, and they are growing and need more nutrients per body weight than do adults. The babies will get supplements the most frequently out of any group of lizards. They should receive the two-thirds vitamin/mineral mixture and one-third calcium mixture on a daily basis (once per day). This means you should dust the insects that will make up one meal once per day.

4. Adults of small species and subadult specimens of larger species can be given the supplement mix two or three times per week. Obviously this can vary, depending on how big the adult specimen is or how close to maturity the subadult individual might be.

P. FREED

Lace-wings (*Neuroptera* sp.) can't be cultured, but can add to the captive lizards diet.

5. Adults of larger species (basilisks, larger chameleons, knight anoles, etc.) will need supplementation approximately one or two times per week.

6. Over-supplementing is not better than following these guidelines.

NUTRITIONAL DISORDERS

Metabolic Bone Disease

This disease is the result of a lack of knowledge or a lack of effort in trying to offer the most nutritionally complete diet possible. This disease is most prevalent in vegetarian reptiles, such as turtles and iguanas, but can occur in carnivorous and omnivorous animals as well. The reason this disease occurs is because there is an improper intake of calcium and phosphorus. The ratio of 1 to 1.5 parts calcium to 1 part phosphorus is extremely important. Any major changes in this ratio diminish the uptake of these necessary minerals. Any hindrance in the uptake of these minerals, especially by growing animals, results in deformity of the bone structure. The bones may actually grow at different rates in different parts of one limb, resulting in bizarre, angled structures that tend to be brittle and apparently very painful.

This condition can be further aggravated by deficiencies in vitamin D. Vitamin D in reptiles takes the form of vitamin D3. With insufficient UV (ultraviolet) radiation and/or vitamin supplementation, the lack of this vitamin can further aggravate a problem with the calcium/ phosphorus ratio. Proper full-spectrum lighting, in the form of some of the fluorescent bulbs available commercially, can assist in the production of vitamin D3.

This sort of ailment should receive a veterinarian's care. The usual therapy for metabolic bone disease is: 1) injections of calcium gluconate two to three times per week; 2) oral vitamin D3, as well as calcium supplementation in the form of calcium lactate or calcium carbonate; and 3) a restructuring of the diet to allow for a more normal entrance of these vital nutrients into the diet of these reptiles. These procedures will not reverse the effects of prior metabolic bone disease and can only prevent it in the future.

B. KAHL

Collared lizards drink infrequently at best. Prolonged drinking oftentimes means a sick animal.

Vitamin A Deficiency

This disease is most prevalent in vegetarian species of lizard, but occasionally insectivores raised in captivity can succumb to the disease. The usual signs are puffy, swollen eyelids, and respiratory distress.

Insects that are not loaded beforehand by the keeper will be

BOTH PHOTOS BY DR. FREDRIC FRYE FROM HIS BOOK *REPTILE CARE*.

Two shots of a Green Iguana suffering from metabolic bone disease receiving injections of calcium gluconate.

nutritionally ineffectual and can lead to a host of problems, one of which is vitamin A deficiency. Obviously the way to reverse this process is to start loading the insects with foods containing vitamin A, such as romaine lettuce, yams, and Swiss chard. Immediate treatment for minor vitamin A deficiency is through oral administration; more severe cases require injection treatments.

Obesity

Some people might consider the possibility that a lizard could suffer from obesity as silly. They might even say that a fat lizard is a happy lizard. Anyway, lizards are very active and should be able to work off any excess fat stores quite easily. It is true that lizards can work off moderate fat stores, provided that their diet is modified. A lizard that is continually fed a diet that is too rich will not get rid of any excess weight; additionally, an obese lizard usually will exhibit symptoms of other ailments as well.

An obese lizard is the result of feeding the wrong foods. An extremely fatty diet, such as one that is mainly waxworms along with crickets that have been gut-loaded with fatty foods, for example, will lead directly to obesity. With moderate obesity, a slight change in diet will slim down a lizard quickly and without harm. However, an extremely obese lizard will need a veterinarian's care in providing extra nutrients, along with a supervised diet in order to have the best chances at avoiding any of the symptoms that can accompany reptile obesity.

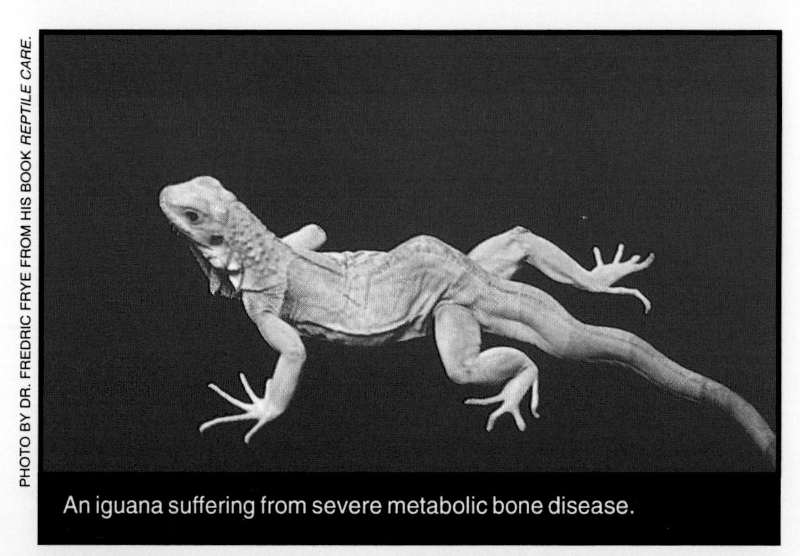

PHOTO BY DR. FREDRIC FRYE FROM HIS BOOK *REPTILE CARE*.

An iguana suffering from severe metabolic bone disease.

Vitamin B Deficiency

This particular deficiency is not known widely in lizard species. The only susceptibility lizards tend to show is with regards to vitamin B2. The usual symptom is paralysis of the hind legs, usually with little or no warning or other symptoms. It seems that iguanids and agamids are the most at risk. The use of a quality vitamin and mineral supplement usually keeps these types of ailments at bay.

Vitamin C Deficiency

An ailment that often is blamed (probably incorrectly) for mouth rot in captive lizards, vitamin C deficiency usually can be reversed with the addition of the vitamin to the normal foodstuffs.

Regurgitation

There are various causes for this. Sometimes it is the result of an inappropriate food item being successfully swallowed that is either too large or contains something that causes the lizard some type of gastric distress. A prey item that is too large will not allow proper digestion to occur, either blocking the stomach or, if partially digested, impacting the intestine.

Another reason lizards regurgitate their food is because of inappropriate temperatures needed for digestion. A lizard that is too cold, with no chance to regulate its own temperature, cannot digest *any* food, let alone large pieces. The lizard either will pass the insect through undigested or will regurgitate it.

Other reasons for vomiting are bacterial and protozoal infestations. These can manifest themselves in many ways including regurgitation, as well as weight loss, diarrhea, etc. The advice of a veterinarian will be needed if you suspect such a cause for regurgitation.

An iguana with a probable vitamin A deficiency. This particular case would need injections of vitamin A, otherwise it would probably lose its sight. Photo by Dr. F. Frye from his book, *Reptile Care.*

Endoparasites

There are several parasites that affect the intestines of lizards. They usually take the form of roundworms or tapeworms, and both can be detected through stool samples. The symptomology includes a good appetite, with no apparent gain in weight. In fact, lizards tend to lose weight while eating well. As the parasites gain hold the lizard starts to show signs of being listless, exhibits a drastic loss of weight, and develops sunken eyes. If left untreated, the lizard usually will perish, no matter how well it otherwise is taken care of. Medication from a veterinarian should be sought immediately.

OFFERING THE INSECTS

Now that you have done the hard part of conditioning and dusting the insects, it would be a shame to have no positive results come from it. An insect thrown into a cage quickly will "make a break for it," trying to hide as speedily as possible. It is definitely worth a little extra time in preparing the insects for your lizards.

A lizard housed in a "naturalistic" setting, i.e., a biotope environment with lots of branches, plants, etc., will be a bit of a challenge to feed. A cricket thrown into the tank will be either quickly gobbled up by the lizard or find a hiding spot before it can be eaten. Once the cricket has found a hiding spot it either may eventually be caught and eaten (without any vitamins or nutrient loading) or it may go uneaten altogether. A mealworm will quickly dig into the substrate and never be seen again. Here are a couple of tips for getting the food to the lizard:

1. For crickets, which should be the usual insects that are vitamin dusted, pinch off the rear "jumping" legs. This can be done by squeezing these legs at the knee. It sounds cruel, but this will allow you to feed your lizards without any insect loss. Once the crickets have been pinched, you can place them in a high-sided bowl placed in the aquarium. The bowl can be buried into the substrate up to its rim to hide it. The bowl should be porcelain, about 2 inches high, with vertical sides; this will not allow the insects to escape.

2. Mealworms, waxworms, and the like can be placed into the same type of bowl as the crickets. Put in only what the inhabitants of the cage will eat in one feeding, though.

3. Winged insects, such as moths and butterflies, can either be "de-winged" before use and put into the bowl or can be allowed to fly around the cage to be chased and caught by the lizards. You will find that lizards love the chase involved in capturing their own insects. The only problem may be that these insects, too, may eventually find hiding spots if not caught immediately.

4. Pinkies should be allowed to stay in the bowl for no more than a few hours before returning to the mother, if you breed mice. If the pinkie was bought from a pet store then this is not a choice. If the pinkie goes uneaten and eventually dies, dispose of it immediately.

FORCE FEEDING

Force feeding occasionally will be needed to be carried out on sick, old, newborn, or improperly housed lizards. This is a drastic step and never should be carried out on animals that currently are feeding by themselves. Force feeding should not be carried out until several possible correctable reasons for the lizard not eating have been reviewed. Check to see if the animal is healthy. To do this you will need to see the veterinarian.

But a veterinarian costs money, so first check the lizard's housing. Check temperature, day/night lighting cycles, possible aggression from other tankmates, etc. If these factors all are acceptable, then a veterinarian is the only solution to the problem. The veterinarian will check out the animal and may clear it or prescribe medication. If the animal is cleared and it *is* being properly housed but still will not feed, then force feeding is the only answer. If the lizard is prescribed a medication, then follow the veterinarian's instructions; if, after the medication course is completed, the lizard still won't eat, force feeding must commence.

The actual force feeding of an animal is extremely stressful, so make sure you understand what was said above. The actual feeding method should be as follows.

1. Feed smaller than normal sized prey to the lizard. This makes digestion easier. Condition the insects just as for healthy lizards. Dust them in the same way, too. Feedings should be done on a slightly less frequent basis than for a healthy lizard, though, because the digestive capabilities of a sick lizard will be hindered.

2. Use a varied diet.

3. A lizard that is picked up may

"gape" at you naturally, but chances are that you will need to open up its mouth manually to feed it.

4. Use a thin piece of plastic or metal to *gently* force open the mouth of your captive. A butter knife (without serrations) or even a credit card will do. Make sure to clean the instrument carefully beforehand, though.

5. Slight pressure on the jaw should be all that is necessary for the animal to open its mouth. Once the mouth is open the prey can be slipped into the mouth. The lizard should then swallow the food on its own. If the lizard spits it out, the prey must be placed in the back of the throat in order for the lizard to facilitate swallowing.

6. *Watch the lizard closely.* If after a minute or so you notice that the lizard is not swallowing, the prey must be quickly removed from the lizard's mouth. If this is the case, liquid food must be given to it and should be obtained from your veterinarian. Liquid foods must be force fed through a tube, a task that can be difficult and even dangerous if a small lizard is involved. Your veterinarian should be able to instruct you in the proper method of tubing your lizard if this becomes necessary.

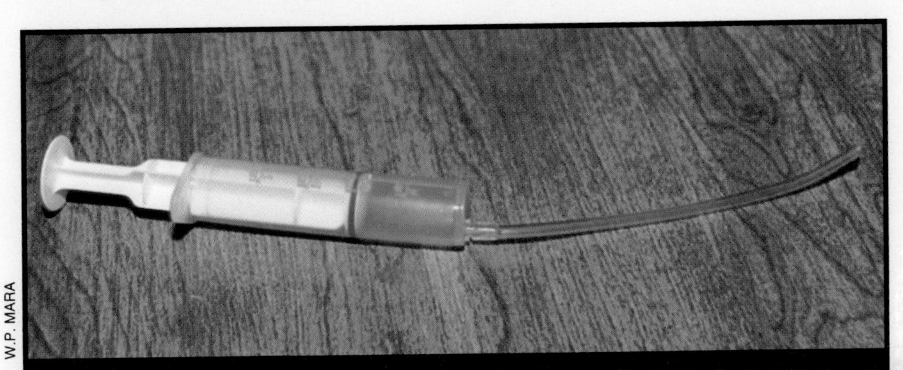

W.P. MARA

A force-feeding tube. This device makes the sometimes difficult job of rehabilitating a sick animal more bearable.

FEEDING HABITS

You very quickly will notice that your lizards will get into a routine of feeding. If they are relatively small they will tend to start to feed every day at approximately the same time (usually once they have warmed up). Larger lizards will not feed every day, but on the days that they do feed it probably will be when they warm up, too. Feeding lizards before they are warm and active usually is a waste of time.

You also soon will notice the particular behavior that goes along with a hungry lizard. Lizards on the prowl are active, sometimes more aggressive than normal, and larger species tend to make a mess of things. They tend to display (i.e., territorial warnings) to each other. Physically, larger lizards that usually don't eat every day will look thinner than normal, and most species will tend to have an altered color pattern compared to their typical coloration. These are just a few tips to assist you in recognizing a hungry lizard.

One way to avoid this is to get the lizard on a regular feeding schedule. This should be done similarly to its supplementation schedule. Here are some guidelines for feeding:

1. Most insect-eating lizards in nature are almost constantly hunting

for food. A Green Anole might eat anywhere from 6 to 20 small insects per day. Obviously it will be impossible to feed a lizard this frequently throughout the day, but, if possible, feed more than once a day for these small, active types of lizards.

2. Larger lizards, such as collared lizards, agamas, etc., are often fed relatively larger prey items. They therefore need more time to digest their food. Lizards such as these can be put on a schedule of feeding that allows you to skip days. Often four or five feedings per week are optimum. Any more feedings and you might have an obese lizard on your hands; any less and you might have a hungry, aggressive, thin lizard. Only experience will be the judge.

3. Newly acquired or sick individuals will eat at a random, sporadic rate. These individuals should be given a diet that is normal for their species. They won't eat at all the feedings, but should be presented with the prey nonetheless (just in case).

4. Many species hibernate in their natural habitats. In captivity, they often will slow down their metabolism or actually go into hibernation, regardless of whether or not there are any changes in the day/night lighting cycles or temperature. Consult specific information on your particular species of lizard to see whether or not this is a common occurrence.

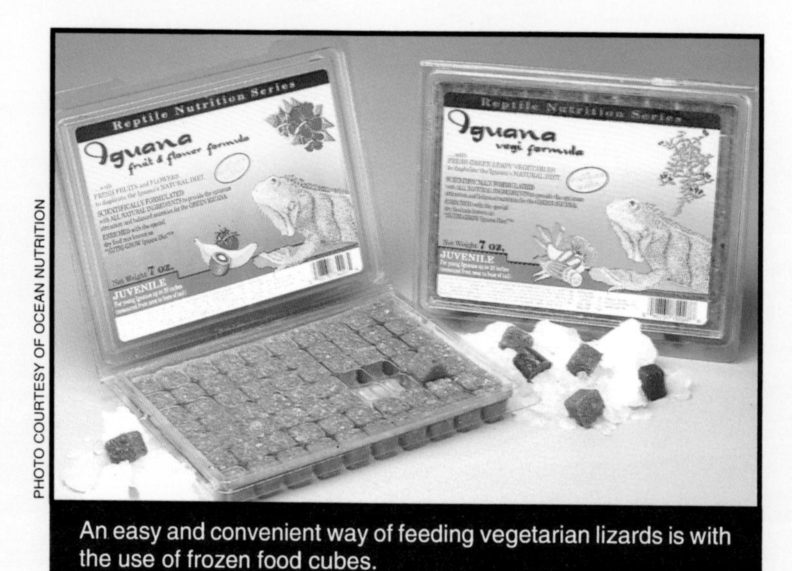

PHOTO COURTESY OF OCEAN NUTRITION

An easy and convenient way of feeding vegetarian lizards is with the use of frozen food cubes.

FRUITS AND VEGGIES

It may come as a surprise to many keepers of "insectivorous" lizards to know that many of these lizards will consume vegetable matter. These lizard species often are opportunistic in nature and will take advantage of these food sources for their nutritional value.

Information on exactly which insectivorous species will take fruits and vegetables is spotty. Offer an assortment of cut up fruits and vegetables in a bowl or plate. It may take several offerings before the lizard will accept it. If after several attempts your lizard won't take the veggies (and it is otherwise eating normally), it is safe to assume that *it ain't eatin' veggies*. Some possible choices for the menu can include: squash, cantaloupe, kiwi, banana, tomato, broccoli, and yam.

On the other side of the spectrum there are the vegetarian lizards that will eat animal matter. These herbivorous lizards also frequently are opportunistic in their natural habitats and will take insects or carrion in their diets as the occasion presents itself. In captivity these lizards, such as most iguanas and water dragons, will take insects quite readily. Try some crickets or waxworms to start. Keep the animal matter in these lizards's diets to under 10-15% of their total intake of food. Check with specific sources on these animals for correct feeding information.

WATER

An often overlooked part of an animals nutrition is water. Many desert species will not need water on a daily basis, but most other types will need to drink clean water every day. All lizards should be provided with clean, fresh water every day, whether they drink it or not. Water should be presented to lizards in a washable container that is in easy sight of the animals. The saying "out of sight, out of mind" works here too. If the lizard can't see the water it won't drink it either. You will tend to notice that water will not evaporate very quickly from a rainforest terrarium. This is no reason to assume that the water will be safe to drink as long as it lasts in the bowl. Tap water generally is safe to use, but even it, if left for several days, will develop bacterial and fungal populations that should not be allowed to enter your lizard.

Rinse out the bowl daily and thoroughly wash it out at least twice a week to prevent "scummy" water from developing. Remember that animals can go many days without eating, but will die within only a few days without clean, potable water.

The Green Anole prefers drinking the droplets from a spray can to drinking directly from a water dish.

M. GILROY

PHOTO OF *PHRYNOSOMA DOUGLASSI ORNATISSIMUM* BY R.D. BARTLETT

Who can resist the cute, cherub-like looks of a horned lizard? You should, unless you are experienced in feeding these specialized eaters.

SPECIES ACCOUNTS

This section is not meant to give basic care information on the various species of lizard, just general information on feeding. The section is set up by groups of lizards, i.e., those that eat specific foods, those that will eat a wide variety of insects, etc. The species in any particular group probably have nothing else in common with each other besides the fact that they have similar dietary requirements.

This section is meant to give you a familiarity with the diets of some of the more commonly seen species in the hobby. Obviously some specimens may not take everything that a particular species is known to eat. Likewise, some specimens may take food items that may not be associated with its species. Use your judgment and common sense to determine what is best for you and your captives and, chances are, you will have no problem.

SPECIALIZED DIETS

Horned Lizards

This is a group of lizards that were once very common in the hobby. They are easily caught species that are cute, small, and look like they would make a perfect pet for a child. But there is a catch. In the wild most species feed strictly on ants and termites and rarely can be weaned onto anything else. These lizards (*Phrynosoma* spp.) often take other prey in captivity, such as fruitflies, small crickets, and mealworms, but it seems that these insects don't get the proper nutrition with this sort of a diet. The possibility that these desert species (North and Central America) are kept at too cool a temperature

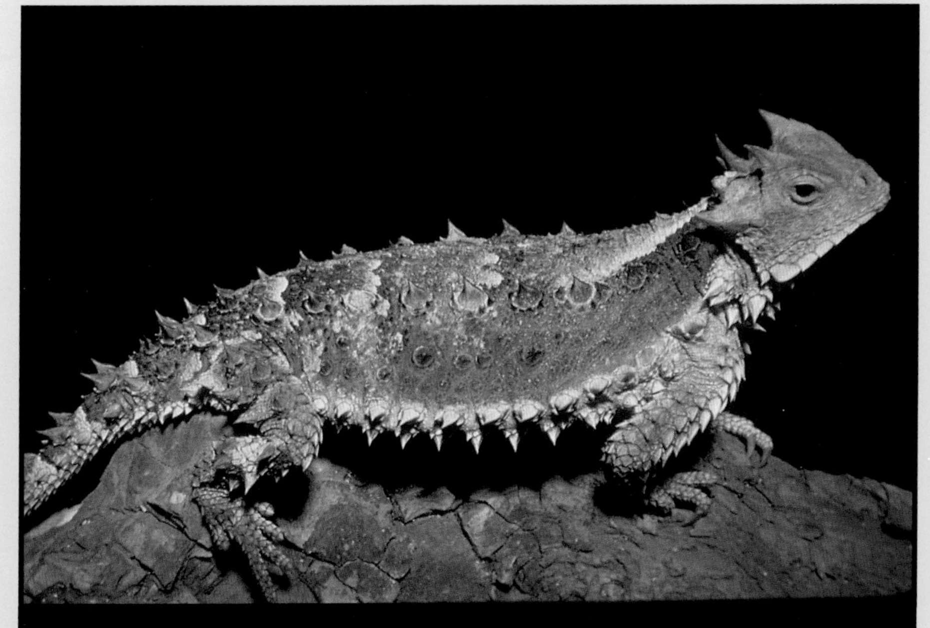

Knowledge of climate, habitat, and natural diet is needed in keeping each species of horned lizard. This is *Phrynosoma asio* by R.S. Simmons.

would mean that they might not be able to properly digest the various foods provided. These lizards should be kept only by experienced herp keepers who can adjust the diets and general housing with some precision to give these lizards the best chance for success.

Sand and Earless Lizards

These lizards of the genus *Holbrookia* are not as specialized as the horned lizards but do require some specific foods nonetheless. Because they usually grow to not more than 7 inches, they require correspondingly small foods. They are not that picky in the prey they will take, but it can become a nuisance to constantly find such small foods. They will take fruitflies, crickets one to two week old, and newly hatched mealworms. Hungry specimens will attempt larger fare, but such prey should be avoided for the overall long-term care of these animals.

Flying Dragons

A group that seldom is seen in the hobby. If it weren't for their specialized diets they would make excellent pets. Their small size and unique physical form are intriguing, but they need a diet of small insects, including ants, baby crickets, and fruitflies. These lizards (*Draco* spp.) are best left to the experienced herpetoculturist.

Moloch

The moloch (*Moloch horridus*) is a species that is seldom seen in the hobby. Its diet is as bizarre as its looks. The reptilian version of a porcupine, thorny projections cover this lizard's body. It eats ants, and in

These earless lizards (*Holbrookia macillata approximatus*) are predators of small insects, including ants, fruitflies, and the like.

R.D. BARTLETT

K. NEMURAS

Above: The Blue-tailed Day Gecko (*Phelsuma cepediana*) needs fruit and/or nectar, along with the small insects that round out it's diet. **Below:** The Australian version of a horned lizard—the Moloch (*Moloch horridus*).

K. LUCAS

fact it eats literally hundreds of them per day. This should divert your attention from this species immediately. A species that is tough to keep for even the experienced herp keeper, it also is protected by Australian wildlife laws.

Ground Skink

Extremely small lizards such as the Ground Skink (*Scincella lateralis*) need extremely small food items. Some acceptable foods include baby crickets, small spiders, and wingless fruitflies.

Day Geckos

These beautiful diurnal (day-active) geckos of the genus *Phelsuma* are becoming extremely popular as pets of late. At one time they were considered virtually unkeepable, but through persistence their dietary needs have been discovered. Day geckos kept solely on an insectivorous diet will waste away and perish. These lizards *need* fruit and nectar in their diets. Fruit baby foods, with added vitamins, tend to work very well, along with fruitflies, small crickets, and waxworms. It is

Facing page: The Flying Dragons (here: *Draco blanfordii*) are Old World lizards that demand small fare on their diets including ants, fruitflies, pinhead crickets, etc. Photo: K.T. Nemuras.

Above: The Ground Skink (*Scincella lateralis*) is an infrequently seen predator of small spiders, crickets, and so forth. Photo: R.T. Zappolorti.

K.H. SWITAK

The highly predacious Collared Lizard (here: *Crotaphytus bicinctores*) eating a Western Whiptail, *Cnemidophorus tigris*.

R.D. BARTLETT

The Dwarf Tegu (*Callopistes maculatus*) feeds on many different types of animals in it's native South America, from spiders to birds.

worth getting some experience with other lizard species before attempting these species.

INSECTIVOROUS/CARNIVOROUS DIETS

Collared Lizards

A group of very closely related lizards of drier parts of North America, collared lizards (*Crotaphytus* spp.) grow to approximately 16 inches maximum but have an enormous appetite. In nature they tend to stay strictly with insect prey, but in captivity they will accept pinkies willingly. They are easy to feed, as they eat virtually anything that moves, often even cannibalizing their own babies. The closely related leopard lizards (*Gambelia wislizenii*) take identical diets.

Tegus

A group of large South American lizards of the family Teiidae. Tegus generally are large, aggressive lizards that, as juveniles, take an insectivorous diet. However, as adults most switch over to a mainly carnivorous diet (small mammals and birds).

Monitors

The young and some of the dwarf species of this genus (*Varanus*) eat a diet of insects. Most species, though, become at least fairly large and quickly outgrow insects for vertebrate foods (from mice to goats).

Basilisks

These large (18-36 inches), arboreal (tree-dwelling) tropical American lizards (*Basiliscus* spp.) feed mainly on insects in the wild. They are opportunistic, however, and are known to take small mammals, baby birds, etc. In captivity they should have a similar diet, although baby birds are not a necessary addition. Crickets can make up the bulk of the diet, with other insects

R. SPRACKLAND

The beautiful Emerald Tree Monitor (*Varanus prasinus*) feeds mainly on insects, spiders, and birds.

and pinkies filling out the rest. Cone-headed lizards (*Laemanctus* spp.) and helmeted iguanas (*Corytophanes* spp.) take similar food items.

OMNIVOROUS DIETS

Rhinoceros and Spiny Iguanas

These large tropical American lizards are true omnivores. They start life as mainly insect-eaters, but quickly switch over to small mammals and vegetation. The spiny iguanas (*Ctenosaura*) tend to take more animal matter than the rhinoceros iguanas (*Cyclura*). The spiny iguanas are much more affordable and available than the rhinoceros iguanas, due in part to the fact that most rhinoceros iguanas are protected by law.

Green Iguanas

The Common or Green Iguana (*Iguana iguana*) usually is listed as a strict vegetarian. This is almost true for the adults, but certainly not true for juveniles. Juvenile iguanas are omnivorous, eating a good proportion of insects along with the vegetation that makes up the rest of the diet. Juveniles that don't get any animal matter in their diets usually are stunted and prone to illness. Iguanas tend to become more and more herbivorous as they grow, so by the time they are adults (3 feet or so) they take only a very small proportion of animal matter (under 10%). Water dragons (*Physignathus* spp.) are unrelated but fill a parallel niche to the New World iguanas, taking similar diets (except that they do tend to take more animal matter).

Blue-Tongued and Pinecone Skinks

These large, mostly Australian skinks (genera *Tiliqua* and *Trachydosaurus*, respectively) are generally thought of as herbivores, but they do best with a fair amount of animal matter in their diets. Being large, (for lizards, anyway) their carnivorous cravings are sated by a variety of prey animals. The blue-tongued skinks are known to eat large insects (locusts, moths, etc.), as well as smaller vertebrates (mice, lizards) along with their usual vegetation. The pinecone skink is likely to take earthworms and snails, with some insects and much vegetable matter.

A. KERSTITCH

Above: This juvenile basilisk (*Basiliscus* sp.) will prosper with a varied diet that has vitamin/ calcium supplements added on a regular basis. **Below:** The omnivorous Water Dragons (*Physignathus concincinus*, pictured here) are Old World counterparts to the iguanids of Central and South America.

R.D. BARTLETT

K.T. SWITAK

The Shingleback or Pinecone Skink (top), *Trachydosaurus rugosus*, and this Blue-tongued Skink, *Tiliqua nigrolutea* are omnivorous and opportunistic feeders that will take a varied assortment of foods.

R.D. BARTLETT

The Ameiva or Junglerunner (*Ameiva ameiva bifrontata*) are tropical, insectivorous lizards that require a large enclosure in order to keep healthy.

INSECTIVOROUS DIETS

Ameivas

Ameiva spp., also known as jungle runners (*Teiidae*), represent a diverse group of lizards native to Central and South America. There they inhabit many habitats, from savannah to rainforest. They all are active, alert, medium-sized (12-24 inches) lizards that, in the wild, consume mainly insect prey. In captivity, when provided with adequate conditions, they tend to be very adaptable about their diets and are known to eat spiders, insects, pinkies, and other lizards (don't house them with your Green Anoles!).

Lacertas

These Old World lizards, native to Europe and eastern Asia as well as Africa, take diets similar to those of the ameivas. They need a base of insects (crickets, mealworms, etc.) to thrive. They will also take, on occasion, pinkies, lizards, and some plant material. Several species of lacertids, often known as jewel lizards, reach the hobby, especially

A male *Lacerta viridis* at home in the open forests that this species calls home. Photo: W. Kirsche.

in Europe, and grow to from 6-30 inches in length.

Fence Swifts and Spiny Lizards

This large group of lizards represents the genus *Sceloporus*. These active, shy lizards of North and Central America are true insectivores. They are too small (6-14 inches) to realistically attempt to eat anything larger than a cricket, and in the wild they feed exclusively on spiders and insects. This should be the basis for their diets in captivity as well. Fence swifts are adaptable, hardy animals suitable for all herpetoculturists, from novice to expert.

Some of the more colorful species of anoles don't make it to the hobby as often as they should, usually because of restricted ranges or protected status.

PHOTO OF *ANOLIS BIMACULATUS* BY D. GREEN

W.B. ALLEN, JR.

Fence swifts (*Sceloporus orcutti*, seen here) are quick, agile lizards of the southern U.S. into Central Mexico.

Anoles

Probably the most common group of lizards encountered in the hobby, the genus *Anolis* (in the broad taxonomic sense) contains the familiar Green Anole or "American chameleon" (*A. carolinensis*), along with over 300 other species that inhabit areas from South America and the Caribbean Islands to the southeastern and midwestern United States. These lizards are excellent animals for the beginner, as they are hardy and adaptable and can absorb a lot of mistakes a novice might inadvertently make. Most anoles are small in size (generally 6-10 inches), but some grow as large as 18 inches (*A. equestris*). The small species eat exclusively insects and small invertebrates. The larger species sometimes will eat small lizards and pinkies. Many of the species also will sample fruits if given the opportunity.

Curly-tailed Lizards

There are several species in this genus (*Leiocephalus*), all which are native to the West Indies (Caribbean), with close allies in South America. There are now even established colonies in Florida. Like many of the iguanids, these lizards are active and aggressive, establishing territories that they defend. In these territories they scour the area for their foods. These medium-sized lizards (8-16 inches) are true insectivores, consuming many sizes and types of insects and other small invertebrates. In captivity the larger species will accept pinkies as well.

Agamas

There currently are over 60 species in the genus *Agama*, though recently several scientists have proposed breaking the genus into several smaller genera. They all are Old World lizards inhabiting Europe, Asia, and Africa. There they fill niches similar to the curly-tailed lizards and collared lizards of the New World. Growing to sizes similar to their New World cousins (6-16 inches), they prey on the same types of insects and invertebrates, as well.

Above: Agamids are a large and varied family that comprises mostly insect-eating lizards, including this *Agama atra knobeli* from Namibia. Photo: P. Freed.
Below: The Red Curly-tailed Lizard (*Leiocephalus schreibersi*) is an active semi-desert/scrubland dweller of the Caribbean. Photo: G. Dingerkus.

R.D. BARTLETT

An adult male Schneider's Skink (*Eumeces s. schneideri*).

Skinks

There are many species of skinks that occur throughout the world. Some grow fairly large, such as the blue-tongued skinks (18-30 inches) of New Guinea and Australia, but the bulk of the skinks reach anywhere from 4-14 inches. Most skinks are similar in appearance (short legs, long tails, glossy scales), and most fill similar niches wherever they occur. This means that their diets and behavior are very similar as well. Skinks generally are very shy, alert lizards that always are close to shelter. They don't bask as frequently as some other lizards and normally can be found in the litter of forests or the sands and rocks of deserts and savannahs. There they eat many of the insects and other small invertebrates that also make the litter their homes. In captivity skinks make good pets, as they can take a wide variety of foods and are very adaptable to conditions offered.

Geckos

Geckos occur throughout the world, but are most common in the tropical and subtropical regions. They are mostly nocturnal predators of insects, invertebrates, and other lizards. Most gecko species (e.g., the Tokay Gecko, *Gekko gecko*, Flying Gecko, *Ptychozoon kuhli*, and the Turkish Gecko, *Hemidactylus turcicus*) have foot pads, known as lamellae, that allow them to climb vertical surfaces in search of prey as well as to elude predators. Some species (such as the Leopard Gecko, *Eublepharis macularius*, and the Banded Gecko, *Coleonyx variegatus*) do not possess such toe pads and are terrestrial in their habits.

The geckos are a family of wide-ranging insectivores that generally are nocturnal, secretive animals of tropical climates. Pictured: *Hemidactylus turcicus*. Photo: K.T. Nemuras.

Above: A Flying Gecko, *Ptychozoon* sp. Their bizarre frills and skin flaps allow it to glide from tree to tree in search of their insect prey. **Below:** The 14 inch Tokay Gecko (*Gekko gecko*) is a familiar sight to herp enthusiasts everywhere.

K.T. NEMURAS

PS-311, 96 pgs, 60+ photos

SK-015, 64 pgs. 40+ photos

PS-316, 128 pgs, 100+ photos

KW-196, 128 pgs, 100+ photos

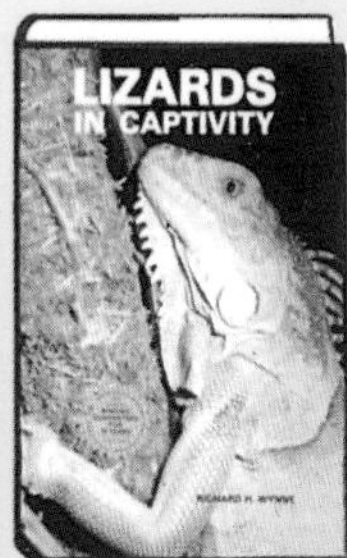

PS-769, 192 pgs, 120+ photos

TU-025, 64 pgs, 60+ photos

SK-032, 64 pgs, 40+ photso

YF-111, 32 pgs

TS-145, 288 pgs, 250+ photos

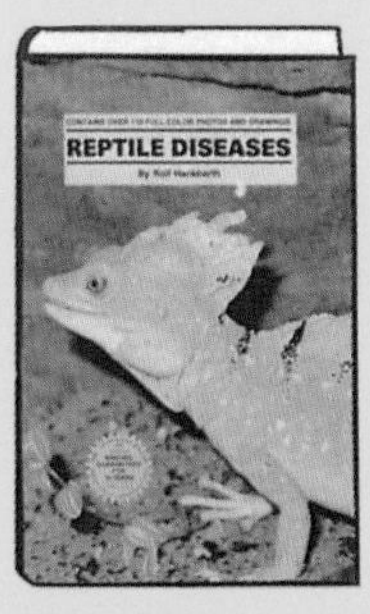

KW-197, 128 pgs, 110+ photos

H-935, 576 pgs, 260+ photos

TS-166, 192 pgs, 170+ photos

H-1102, 830 pgs, 1800+ Illus and photos

TS-165, VOL I, 655 pgs, 1850+ photos

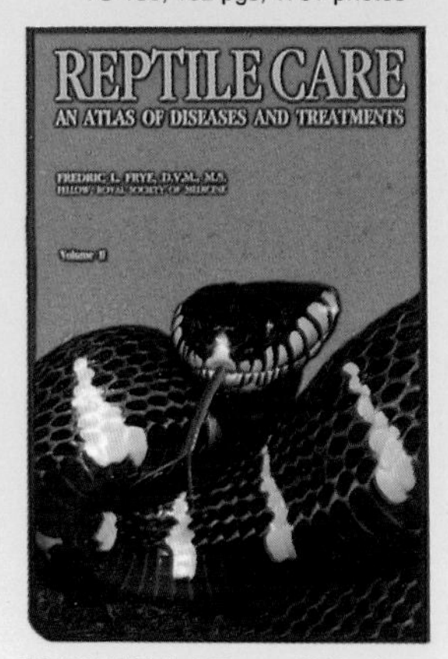

TS-165, VOL II, 655 pgs, 1850+ photos